MY little BIG MATH BOOK

LARS RÖNNBÄCK

illustrated by
LÍDIA STEINER

To my wife Anneli,
and our daughter Nathalie.

Published by UP TO CHANGE,
Karlbergsvägen 67, 113 35 Stockholm, Sweden.
http://www.uptochange.com

The running text in the book is lovingly typeset in Sina Nova made by Dieter Hofrichter with headings in Source Sans Pro by Paul D. Hunt. The cover title is typeset using You Blockhead by John Roshell and Gabriola by John Hudson.

Parts of the book are based on research from ANCHOR MODELING, http://www.anchormodeling.com.

ISBN: 978-91-982826-0-3 (revision2)

Preface

The goal of this book is to through simple illustrative scenarios describe the early mathematical development of children and how they become aquatinted with notions usually reserved for studies in higher mathematics. The book will also give an understanding of the general order in which this mathematical knowledge is gained.

Most children have gained the knowledge described in the book before turning eight, and most of them through experimentation driven by their own curiosity. Very few of them have reflected upon what they have learnt through these experiments though, having to familiarise themselves with these concepts again much later in life. Because much of the described knowledge can be seen as self-evident, it is easy to overlook its connection to the foundations of mathematics, and near impossible for child to figure out alone. With this book and some encouragement, the scenarios can be discussed and experiments reenacted together with an adult, in order to give a child a deeper and objective understanding of the concepts involved. Such an understanding will undoubtedly help and better prepare children for their journey through the mathematical landscape of life.

The child depicted in the scenarios age as the book progresses, but no exact age is given for when each scenario is likely to occur. This is a deliberate choice. Since the development of a child is individual, children may reach these scenarios at different ages, sometimes in a different order than the one presented, and sometimes in parallel with others.

In order to give children, parents, teachers, and other curious minded people reading the book some background, each scenario is accompanied by a short section explaining the underlying mathematics together with suggestions for further experiments. For those interested in finding more information the words in *italics* in each section provide good search words to use on the Internet. I hope reading the book will be an enlightening adventure for everyone involved!

I would like to thank my family for their endless motivation, Lídia Steiner for her fantastic illustrations, and Camilla Björklund, associate professor at the Department of Education, Communication and Learning, University of Gothenburg for reading and providing thoughtful comments on the manuscript.

One day I discovered something else.

The mathematical journey of a child starts early, maybe already in the tummy, when an unborn child becomes aware of its own existence. Within the area of *logic* you can find the *existential quantifier*, through which you simply can state that something exists, which is a necessary requirement for making a thing an object of further enquiry. In doing so, you have also indirectly given this thing that exists an *identity*. You single out this thing as being something different from everything else. It would be impossible to talk about something specific if there was only one single thing, because then there would be nothing against which you could differentiate it. *Existence* and *identification* are two very important and fundamental concepts in mathematics. The first contact with these concepts for children is when their sense of *self* is formed, thanks to the discovery that there exists something outside of themselves. Identity is an abstract property that you assign to a thing, special in the way that it can last even if all other properties of the thing change. When one thing transforms into a different thing is a subjective matter, since it must be decided by the one who assigned the identity in the first place, who then would have to assign a new identity to the different thing.

As an experiment, think of what it would take for a few things around you in order for you to decide that they had become something else. Is there something that is hard to determine whether it exists or not?

Another day I found something different.

A child quickly learns how to identify other things. It is no longer just *the self* and *the else*, where the else now seems to consist of many distinguishable things. In order to tell two things apart there has to be at least one *property* in which they differ. At the very least they must occupy different positions in *space*, if being identical in all other aspects. Things that occupy the same place at the same time must be the same thing. Things that are in different places at the same time cannot be the same thing. But also differences differ. They can be the described *locus*, where things are placed, or *features*, inherent properties of things such as colour and shape, *peripherals*, things related or connected to things, *census*, what others say about things, and *heritage*, where things came from. All of which a child uses to *differentiate* things. A blue bed lamp is easy to tell apart from a red ball. Therefore, if you want to encourage the ability to identify things it helps to surround a child with things that have many differences. Most toys do, so it is rarely a problem. Mathematically, the ability to distinguish many things provides the basis for *set theory*, in which you collect various things into sets.

As an experiment, find two things in your surroundings and describe all the differences you can find between the two. Can you find two things that are hard to tell apart?

Then came a day when I found the same thing again.

The notion of *sameness* does not only require the abilities to assign identities and tell things apart, but also that you have a *memory*. Since the identity of a thing is an abstract property, it is not what is remembered. Instead, the identity can be seen as an address to where you have stored the memory of other properties of the thing. When recalling a thing, some, and not all, of these properties normally suffice in order to conclude that this is something you have seen before. Such combinations of properties are called *identifiers* for the thing. Your brain will then connect these to provide a shortcut to the address to where you can find the rest of the memory. The efficiency of this process is quite evident to anyone who has attempted to substitute an old favourite cuddly animal for a new seemingly identical one. When a child starts to recollect things it also gets a feel for time and space, since the same things will suddenly turn up elsewhere, causing wonder of how and why this *temporal* and *spatial* movement occurs. It may be argued that for a small child it is easier to learn how to recall a few things rather than a plentitude. Mathematically, sameness is of great importance in *set theory* since the same thing can only exist once in any given set.

As an experiment, try to find something you know is the same thing, but hard to recognise as such. A photo of a grandmother when she was young perhaps? Think about how you know it is the same thing, and how you could know for sure.

After a while I could recall many different things.

Children will have a perception of *sets* long before they can count. They can see a number of things in a single glance, recalling them or assigning them new identities, distinguishing them from each other. Instead of focusing on *one* thing at a time they can now see and conclude that there are *many* things. The most basic way of creating a set in set theory is through *extension*, simply listing its *members*, such that the end result is a list of identities in no particular order. The set remains the same as long as it contains the same things regardless of their *order*, and an identity can only occur once in the list of members. For example, if Teddy and Bunny are different things they constitute the same set as Bunny and Teddy. Duplicates are not allowed in sets. Two lists containing the same things but in different order are however two different sequences, and a sequence may also contain the same thing more than once. Duplicates are allowed in sequences. You need to know both the members and their order if you are going to tell two sequences apart, but only the members for two sets. Distinguishing between one and many forms the basis of *number theory*, ordering is a fundamental concept in *order theory,* and sequences are common in *computer science*.

As an experiment, try to find some things where a set is enough to describe them as a collection and some things that require a sequence. Can you find a sequence where order does not matter?

There were fewer things here than over there.

All *finite* sets have a property called *cardinality*, corresponding to the number of members in a set. Even before children can use counting to determine this number, they have an intuitive understanding of *greater* and *fewer*. This understanding can be used to order sets with different cardinalities. A child that cannot count can still determine where you can find the greatest number of things, provided that the *relative difference* is large enough. The *absolute difference* in cardinality between a couple of sets having one and two members is one, but the relative difference is that one set is two times larger than the other. The absolute difference between a couple of sets with ten and eleven members is also one, but the relative difference is that one set is only one tenth larger than the other. Cardinality is unaffected by the individual properties of the members, such as size and weight. Actually, five ants are greater than four elephants, even if the elephants clearly weigh more and take up more space than the ants. Greater and fewer are important concepts in *number theory*, *group theory*, and the study of *mathematical functions* in *calculus*.

As an experiment, take a few sets of things and order them by their cardinality, then look at them both forwards and backwards to get an idea of how increasing and decreasing orders are related. How big a relative difference do you need in order to easily determine which set is larger?

Some of the things were similar.

After a while of being just things, identifiable through their differences, *similarities* are discovered. Through their similarities children can begin to *categorise* things. Some of the things may have the same colour but different shapes, while some other things have the same shape but different colours. Since no two things are exactly alike, being similar translates to being *approximately* alike in some aspect. While none of a soccer ball, an orange, a pea, and the sun are perfectly spherical objects, they may be classified together as all being round things. While two children may disagree on the roundness of an orange, *definitions* in mathematics are *precise*, so that we know exactly how much an orange would have to be squashed before it stops being round. Two mathematicians will therefore always agree on the roundness of a particular orange. *Unambiguousness*, that an orange cannot be flat and round at the same time, is an important concept in almost all scientific research. *Measure theory*, *statistics*, and *approximation methods* are some of the areas in which similarities play an important role.

As an experiment, examine two things and describe the similarities they have. If they are only slightly similar in some aspect, how much more different would they need to become to stop being similar? Can you find two things that have absolutely no similarities?

I could put similar things together.

Once children can find similarities such that similar things can be collected they have the tools for creating more advanced *sets*. In addition to creation by extension, sets can now also be created by *intension*, which instead of listing things collects them by one or more *properties* in which they are similar. The set of everything that is green and the set of everything that is round will both contain peas, but soccer balls are only round. After all, a green soccer ball would not be very practical. The set of everything green is also related to the set of everything that has a colour. There are many colours, of which green is merely one, which mathematically corresponds to green things being a *subset* of coloured things. Any member of the set of green things must also be a member of the set of coloured things. Being green is therefore a *restriction* of the more generally applicable notion of having a colour. When things share a number of properties, those in themselves may constitute a named *type*. A thing must have some very particular properties to belong to the pea type, for example. A pea type is also a *specialisation* of a seed type and a seed type is a *generalisation* of peas, beans, rice, nuts, and so on. Through the collection of numerous things, children will rapidly discover new properties from which sets can be created. With this knowledge most of the basic notions in *set theory* have been covered.

As an experiment, first decide a few properties, then try to find everything around you that have them either alone or in combination. Can you find a property that is a restriction of another?

Similar things could get lost among themselves.

When many things become too similar the concept of identity is easily lost. Given a hundred peas in a package, if you pick one up, study it carefully, and then put it back in, you will still have a very hard time finding it again. I could possibly also trick you into believing that I have one hundred peas, even if I have only one, showing you one pea at a time but using the same pea every time. For a child on its way to start *counting*, it is therefore inconvenient to start with things that have very few differences. The primary prerequisite for counting is *identification*, and identification is easier of things that have many differences. That being said, mixing up similar things is not necessarily a bad thing, since it will sharpen the identification skill. If the things to be counted share at least some similarities, the child will have two options for set building, either intension or extension, of which one may be more comfortable than the other. The distinction between being the same thing and being a similar thing is vital in all of mathematics. The concept of *equality* is defined in *logic*, for which when distinction becomes uncertain has a branch called *fuzzy logic*.

As an experiment, try to figure out a way to manipulate the hundred peas so that finding one again would be easier. Can you find other things that can get lost among themselves?

I could split the same pile in different ways.

The same things can be members of many different sets. Given almost any pile of things, it can be divided in numerous ways according to their properties. Say that a pile has both round and green things, from which you can form a set of all the green things and a set of all the round things. If there was a pea in the pile, it should be a member of both the sets, since it is both green and round. This situation provides an excellent opportunity to familiarise yourself with two operations that can be performed on sets. The *union* of those two sets is everything that is green *or* round. The *intersection* of those two sets is everything that is green *and* round. If the pea was the only thing that was both green and round, it is in the union, once, together with all the other green or round things, but it is alone in the intersection. The union, additional properties where things only have to fulfil one, provides more allowing conditions. The intersection, additional properties where things have to fulfil all of them, provides more denying conditions. With the two operations, children now have a rather good grasp of *set theory* and the corresponding *logical connectives*, and and or.

As an experiment, take a pile of things and divide it into, possibly overlapping, sets. Can you find sets with an empty intersection? Are the sets you made, their intersections or unions, subsets of each other?

Even though I took a thing from a pile
there were many left.

In order to learn how to *count* in general, a child must first be able to count to *one*. Children become aware of the fact that one is fewer than *many* at an early age and often use these terms to describe sets of things. There is one pea here and many peas there. In fact, it is possible to repeatedly take one pea from a pile of many peas. Every time you take a pea, there will be fewer peas left in the pile, and the number of peas in the pile will have changed. This revelation begs the question of how many times you can do this before the pile is exhausted. If you place each of the taken peas in a new pile you will eventually have reversed the situation, when only one pea is left from the original pile. Now there are many peas here and one pea there. The one corresponds to that which is *bounded* by a single identity. It could be one house, but also one apartment in that house, or even one window in that apartment, even if they are sequentially *encapsulated*. There is one house over there with many windows. *Peano's axiom* in logic deals with how to repeatedly add one to build higher numbers. Taking and placing one thing is also the first time a child sees *subtraction* and *addition* work.

As an experiment, look around and find something of which there is one in one place and many of in another. Can you think of other ways to exhaust a pile of things?

The pile was gone after I took all the things.

Two other important concepts that children learn are *everything* and *nothing*. There is nothing left if you take everything and everything is left if nothing is taken. However, taking nothing from nothing leaves nothing and nothing is also left if you take everything from everything. This nothing is called the *empty set* in *set theory*. All empty sets are the same, in the sense that they have the same members, precisely none. The empty set is also a subset of every set. Since this set has no members, its cardinality is the number *zero*. The taking of nothing, leaving you empty handed, therefore corresponds to *subtraction* with the number *zero*. Seen the other way around, you can also add nothing to a pile as many times you like without affecting the pile, corresponding to *addition* with the number zero. The everything is trickier, since the everything is *contextual* and always relates to some particular set, with its corresponding cardinality. The derived rule would be that subtracting the same number from itself always yields zero. These rules lay the foundation for *arithmetics*, where addition and subtraction of zero is called an *idempotent operation*.

As an experiment, take everything and nothing from a pile of things. Can you also take everything from nothing, and what would everything be in that case? If you take a set and add it to itself, adding everything to everything, what would the resulting set look like?

Lining up things from different piles
I saw which one had the most.

One way to determine which pile has the most things, which also works well for piles having almost the same number of things, is to *line up* things in rows. Then the pile yielding the longest row will have the most things. The difference in cardinality, which can be hard to determine by just looking at piles without arrangement, becomes obvious once piles are lined up. Arrangements that form some kind of *pattern*, like lines, make it much easier to determine the *relative cardinalities* of sets, even if the exact numbers are not known. Many, but not all things suit themselves to being lined up. Differing vastly in size or enjoying mobility, such as a number of gigantic still-standing skyscrapers and a number of minuscule antsy ants, can make it more difficult. When children start lining things up they are showing an interest in how many things they have. The rows represent *injective functions*, or *bijections* if the sets have the same cardinality. This is also a refinement of the intuitional sense for fewer and greater, since with this method even large sets whose cardinality only differ by one can be distinguished.

As an experiment, take a couple of piles for which it is hard to determine who has the most things and line them up in two rows. Which pile had the greatest and which had the fewest number of things?

There were other patterns I could make from the piles.

If there are many things in the piles the resulting long lines tend to become cumbersome to manage. It is then easier to break a line at some point and start a new parallel line next to it. The pattern can actually be arranged in almost any fashion as long as the result gives you the necessary *visual cues,* but some patterns have turned out to be more practical or common than others. A square pattern saves space, but relies on an *estimate* of the greatest number of things. A three-by-three pattern is only good as long as there are no more than nine things. As long as the pattern is filled according to the same rule for every pile, a single glance will be enough to determine the relative cardinalities. The disadvantage of patterns over single lines is that things may vary even less in size and mobility before the ability to do *pattern recognition* is impaired. Comparing two ants and two elephants to three ants and two elephants may have you lose sight of the ants. The more similar the things are in shape and size, the better the *discernibility* of the patterns become. As a result, the peas may come in handy here. First, place one pea at each thing, then compare the two piles of peas instead. The peas have formed a *one-to-one mapping* with the ants and elephants. Breaking the lines is a step towards a *number system.*

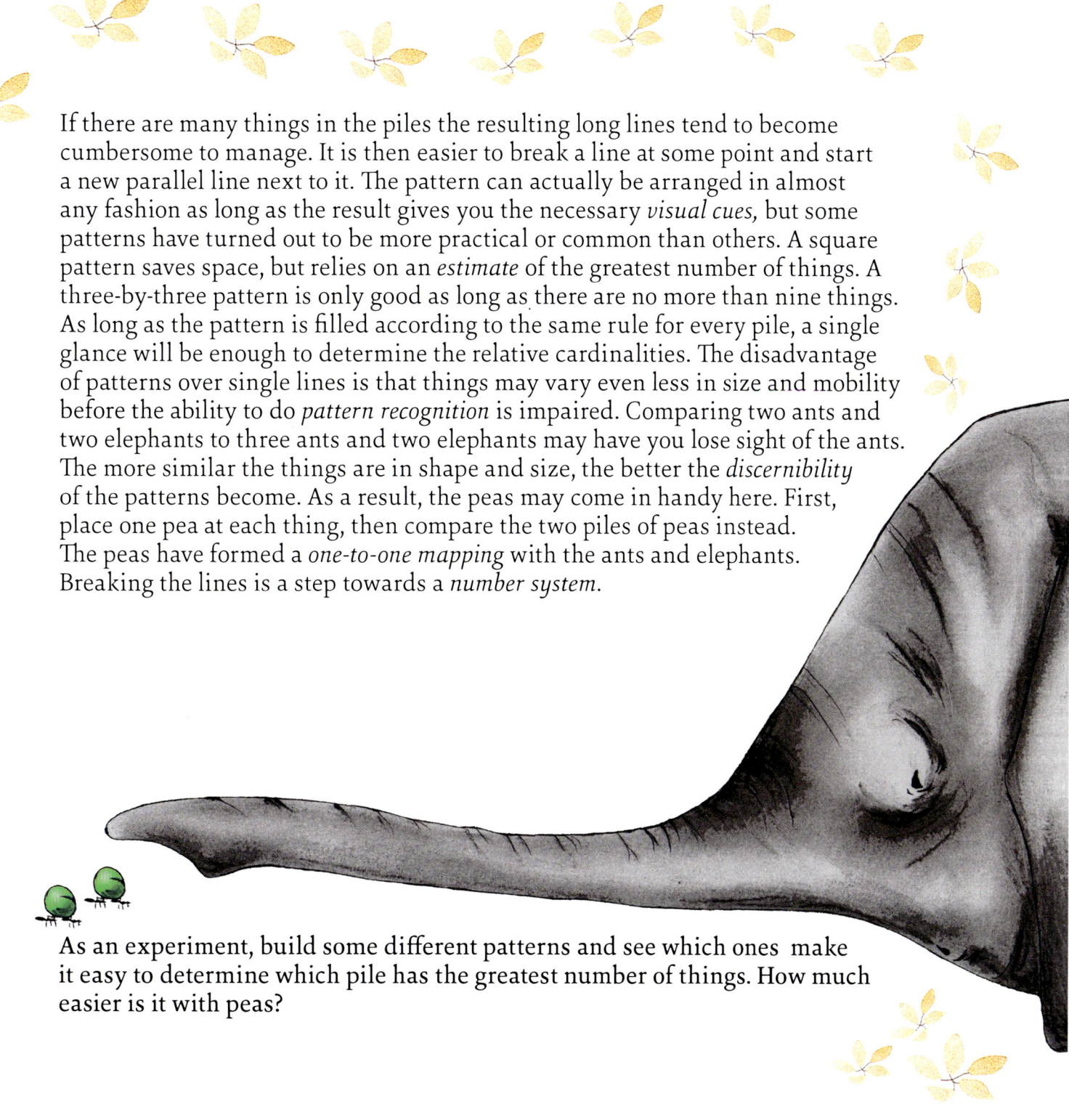

As an experiment, build some different patterns and see which ones make it easy to determine which pile has the greatest number of things. How much easier is it with peas?

I learnt that numbers can be names of patterns.

Once some rule has been established for how to form a pattern and children have familiarised themselves with filling it, the pattern itself can be envisioned and remembered. The *cardinality* of a set of things that would fill such a pattern can then become a name for the pattern. The pattern is the same even if the things change, as long as you have the same *number* of things. One such familiar set of patterns are those found on a dice, with six out of a possible nine positions being used. Take the number five, for example, which can be illustrated using the dot pattern of a dice, spoken as the word five, and symbolised by the digit 5. All of these are valid *representations* of the same number. Having one pea signify one thing, yielding a *one-to-one mapping* between peas and the things to be counted preserves cardinality. For this reason it is possible to use such an exercise to simplify pattern building. In the mathematical terminology, cardinality is *invariant* under one-to-one mappings. A *permutation* is a rearrangement of the same things, but since it is always possible to create a one-to-one mapping between two such arrangements, the cardinality will remain the same. Invariants are an important concept in *topology* and mappings form the basis of *linear algebra*.

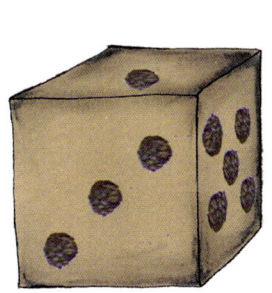

As an experiment, figure out a way to extend the three-by-three pattern of a dice to represent the numbers from six up to nine. How many ways to represent the number five can you find? Do you know of any other patterns that are normally associated with numbers?

I could count by building patterns one thing at a time.

The reason most of us cannot simply look at a reasonably large pile and tell how many things are piled up is that a set is *unordered*, while numbers are *ordered*. When children are counting, they are effectively applying an order on the set, ordering the things into a pattern one thing at a time and remembering the name of each pattern. Sometimes the things you are *counting* are inconvenient to place into patterns, like houses on a street, and you may not have any peas with you to create a *one-to-one mapping* with. Fortunately, if the number is not too great, we can map to our fingers instead. Even before learning how individual numbers are ordered, it is possible to count to ten using this method, as long as children recognise the patterns up to two full hands, having names from one to ten. The better a child gets at remembering the order of numbers, the fewer are the times when fingers need to be used. You only need to remember the last number you counted to and then pick the next one in the sequence when you pass the next house on the street. One becomes two, two becomes three, and three becomes four, and so on. *Numbers* are the *operands* of *arithmetics* and counting can be used to understand *operations*, such as *addition*, on them.

As an experiment, count the number of things in a few piles. If you change the order in which you count the things, does the number remain the same? Are there things around you where you can tell how many there are without counting? Can you find something that is hard to count?

Some of my piles had the same number of things.

Numbers take on a life of their own when children realise that it does not matter whether you count houses, teddy bears, peas, or fingers. The number five is a property of very many sets, namely all the sets that have exactly five things in them, regardless of what those things are. This makes it possible to talk about five generally, without referring to any particular set of things. *Arithmetic operations*, like *addition* and *subtraction*, are *generic* and do not depend on whether it is houses, teddy bears, peas, or fingers that take part in the operations. By removing the connection to specific physical sets of things, numbers are raised to another level of *abstraction*. For children, taking this step can become quite a leap, and it may take many jumps between the levels before they feel comfortable. There is a big conceptual difference between asking how much 5 + 5 is and what one handful of fingers together with another handful of fingers is. The key to arithmetics is to be able to separate the *physical* and *abstract* notions of *numbers*. Being able to make abstractions is also necessary to understand *algebra*, which introduce *formulas* with *variables* that hold *true* if replaced by some numbers, such that a + b = b + a for all numbers a and b.

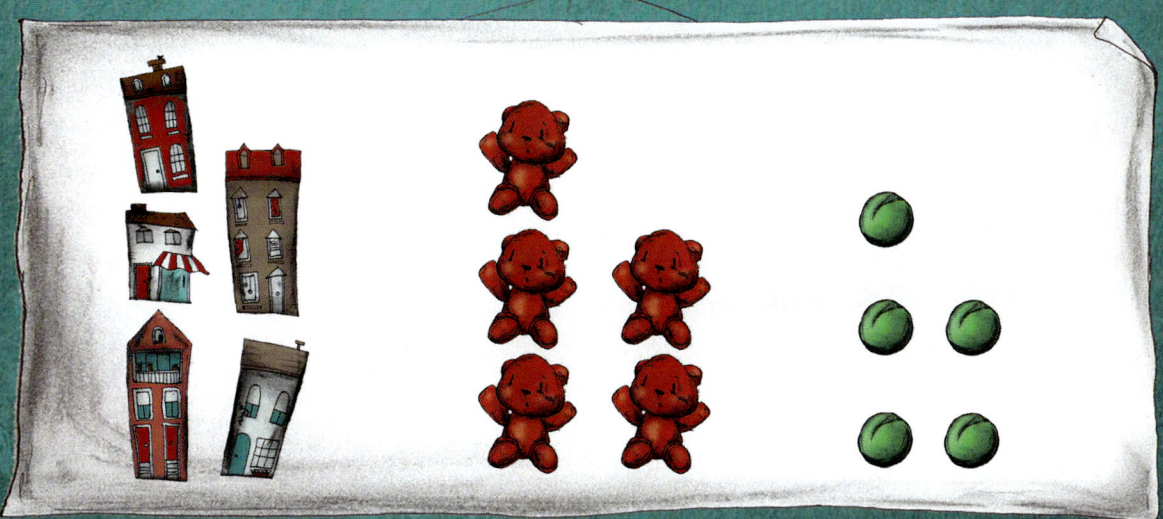

As an experiment, find every set around you that have five things in them. Are there more sets with three things in them than sets with five things in them?

I got a new thing by trading four of my old things.

Denomination occurs when a child decides that one thing is worth more than some other thing. For example, a teddy bear could be worth four oranges, and an orange nine peas. So, with 36 peas you would be able to get the teddy. Through such a *chain of denominations, multiplication* is discovered, and in this case that four times nine is 36. Denominations tend to occur naturally when physical things have different *values*, and even before there were *currencies* trading could be done using chains of denominations. Currencies greatly simplify trading, since all values can be described with just one step in a chain, the *numerical value* of the thing in the selected currency. A currency in itself may be connected to something of value, like gold, but it is more common today to determine denomination by how differently valued things relate to each other. When such relationships change, the currency *inflates* or *deflates*. The idea that things have value and can be traded for other things brings the child closer to the encoding of numbers in *positional notation*, where the position of each digit represent *levels* in the chain of denomination. For example, the number thirty-six is written as 36.

As an experiment, think about the different values of things around you, and if you can trade some of the ones you see for many of the others. Can you produce a chain of denominations?

I put price tags on several of my things.

The patterns we have worked with so far only work well up to a certain size before becoming unmanageable. Expensive things would get very large price tags if only dots are used, and it would take a long time to determine if we have enough things to pay for another thing, putting one thing on every dot until the price tag is filled. The size of the tag can be dramatically reduced if it is split into different parts, such that if the pattern in one part is full the next dot goes into the other part and the full pattern is emptied. Two dice-like patterns, can suddenly represent the numbers between zero and 99, compared to zero to 19 without positional value. With three patterns, the numbers can go up to 999, compared to 29. We save an *exponential* amount of space using this technique. Similarly, modern price tags use *numbers*, where each *digit* in the number determines the value of that digit. For the number 36, the digit 3 in the second position from the right denotes a value of three tens in the first position. The number 123 is one hundred and twenty-three, not one, two, and three. Both the three-by-three pattern and digits use the *base* 10, since both can represent ten numbers, from zero to nine, at the most. Other bases can be used, such as 2, which is commonly called *binary* representation, using only ones and zeros.

As an experiment, make some price tags for things around you, using a base and symbols of your own choice. Do you have a hundred somethings around you? A thousand? A million?

I tried to make a pile of everything red, but had to settle for things in the room.

When sets are created using *intension*, from some *property* the things should share, they can become quite extensive. Imagine the set of everything red. From the things in a single room it may be possible to pile them all up, but if you also look in the next room or out your window there soon will be more things than you can handle. It doesn't stop there either, should you manage to find the red things in the house and in your yard, you need to continue to the next house and the next yard, and so on through the deepest woods to the outskirts of our universe. It is, therefore, often not practical to talk about sets without some *context* into which they fit that also *limits* the things to choose from. There are two ways that this can be done. Either a *universal set* is introduced, containing the things valid for *inclusion* in any set, or the defining property is restricted from "everything red" to "everything red in the room". In *mathematical constructivism*, context and universal sets play an important role, since only sets for which there is a *finite process* to create them exist.

As an experiment, take some rather general property and use it to create a set, then limit this until you only have one thing left in the set. What is the largest set you can think of?

Today I wondered how many things there are.

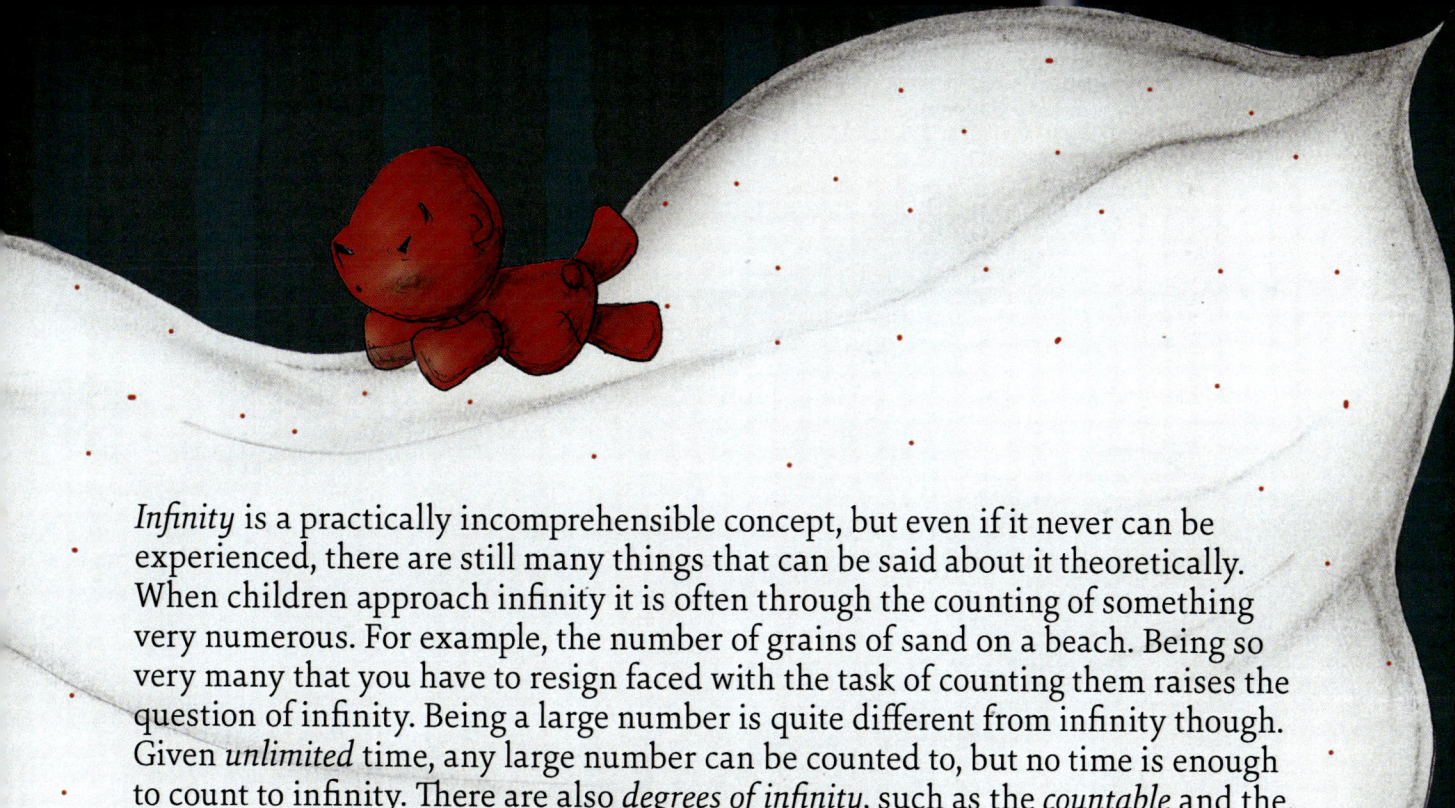

Infinity is a practically incomprehensible concept, but even if it never can be experienced, there are still many things that can be said about it theoretically. When children approach infinity it is often through the counting of something very numerous. For example, the number of grains of sand on a beach. Being so very many that you have to resign faced with the task of counting them raises the question of infinity. Being a large number is quite different from infinity though. Given *unlimited* time, any large number can be counted to, but no time is enough to count to infinity. There are also *degrees of infinity*, such as the *countable* and the, in some respect larger, *uncountable* infinity. The *continuum*, such as the number of points on a line, is the first degree of infinity above the countable. For any two points on the line you decide to count it is always possible to find a point in between the two that you forgot. *Numbers* can be extended to *cardinals*, where higher cardinals denote rising degrees of infinity.

As an experiment, try to find some set that is infinite but still countable. Can you also find a set that is infinite and uncountable?

Tomorrow I will think about if things
I think about also can be counted.

Coming back to the *identities* with which we started, they can not only be assigned to things piled up in front of you, but also to that which we think of, or others have thought of before us. As long as it can be given an identity, it can be counted. Laying in bed counting sheep jumping over a fence it is actually not the sheep we are counting, but the number of jumps. If the same sheep jumps over the fence again, we would keep counting. However, there is even little need to single out any particular jump, so in fact we are really only reciting the *order of the numbers*. There is no need to *recall* a specific jump, which separates them from that which we want to *remember*. If you ask a child how many friends they have, they will most likely be able to count them without them being present, but counting things from *memory* alone may be difficult, and it could happen that the same friend is counted more than once. However, if the names of the friends are first written down, the task of counting them is greatly simplified. The *abstract* thought is *concretised* by *physically* naming them, and concrete things can be *visualised* and *mapped* to *patterns* much more easily than the abstract things in our minds.

As an experiment, try to count something that is not in your vicinity. Can you count something transient, such as how many times you blink or breathe in a minute?

Printed by BoD˝in Norderstedt, Germany